Rhinos

Camilla de la Bédoyère

Quarto is the authority on a wide range of topics.
Quarto educates, entertains and enriches the lives of our readers—enthusiasts and lovers of hands-on living.
www.quartoknows.com

Author: Camilla de la Bédoyère
Editorial: Emily Pither
Design: Cloud King Creative

© 2020 Quarto Publishing plc

First published in 2020 by QEB Publishing,
an imprint of The Quarto Group.
26391 Crown Valley Parkway, Suite 220
Mission Viejo, CA 92691, USA
T: +1 949 380 7510
F: +1 949 380 7575
www.QuartoKnows.com

All rights reserved. No part of this publication may be reproduced, stored in a retrieval system, or transmitted in any form or by any means, electronic, mechanical, photocopying, recording, or otherwise, without the prior permission of the publisher, nor be otherwise circulated in any form of binding or cover other than that in which it is published and without a similar condition being imposed on the subsequent purchaser.

A CIP record for this book is available from the Library of Congress.

ISBN 978-0-7112-5334-6

Manufactured in Shenzhen, China HH102019
9 8 7 6 5 4 3 2 1

Photo Acknowledgments

Alamy: 31 Images of Africa Photobank, 16 John Bennet; **Ardea:** 11t; **Getty:** 13t Doug Cheeseman, 14, 23b Martin Harvey, 15 Whitworth Images, 19, 25 Mark Carwardine, 23t Christian Heinrich; **National Geographic Image Collection:** 7b, 10; **National Geographic Image Collection/Alamy Collection:** 27; **Nature Picture Library:** 4-5, 11b, 13b Mark Carwardine, 21 Denis-Huot, 23b Ann & Steve Toon; **Robert Harding:** 8b; **Science Photo Library:** 28 Peter Chadwick; **Shutterstock:** 1 Handoko Ramawidjaya Bumi, 2 Gualberto Becerra, 6 Michel Arnault, 7t Reinhold Leitner, 8t Doug Cheeseman, 9 Matyas Rehak, 12 Simon Eeman, 17 Travel Stock, 18 Ksenia Ragozina, 22 Volodymyr Burdiak, 24 Lynn Yeh, 26 FotoMiguelRomero, 29 Rich Carey, 32 Noradoa.

Contents

What is a rhinoceros?	4
Jungle rhinos	6
Grassland rhinos	8
What do rhinos look like?	10
What do rhinos eat?	12
In the mud	14
Look, listen, smell	16
Jungle life	18
Herds	20
Baby rhinos	22
Growing up	24
On the move	26
Danger!	28
Saving rhinos	30
Glossary	32

What is a rhinoceros?

A rhinoceros is a big animal with four short legs. It has one or two horns on its nose.

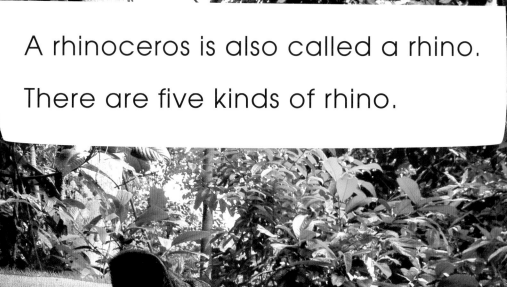

A rhinoceros is also called a rhino. There are five kinds of rhino.

This rhino is 10 feet long.

Jungle rhinos

Two kinds of rhino live in the **jungle**. They are Sumatran rhinos and Javan rhinos.

Jungle rhinos live in the **rainforests** of Asia.

This is a Javan rhino.
It has one horn.

A Sumatran rhino is smaller
and it has more hair.

Jungle rhinos have dark gray or brown skin.

Grassland rhinos

The other types of rhino live on **grasslands** and in **woods**.

This is a white rhino. It is the biggest rhino.

This black rhino likes to roll in mud.

Indian rhinos have thick skin, with lumps, bumps, and folds.

An Indian rhino has one horn.

what do rhinos look like?

Rhinos have one or two horns.

A rhino has thick skin and a big, heavy body.

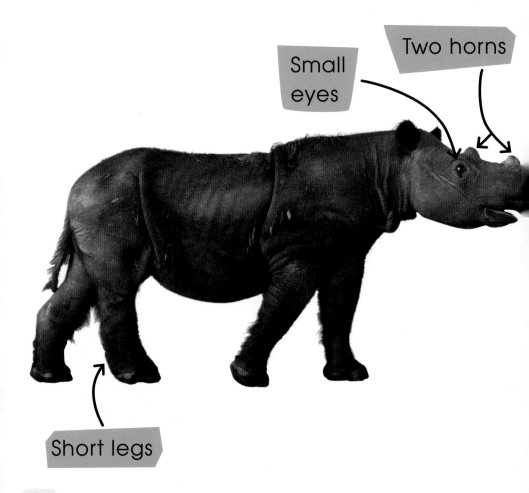

Many rhinos have long hair on their ears.

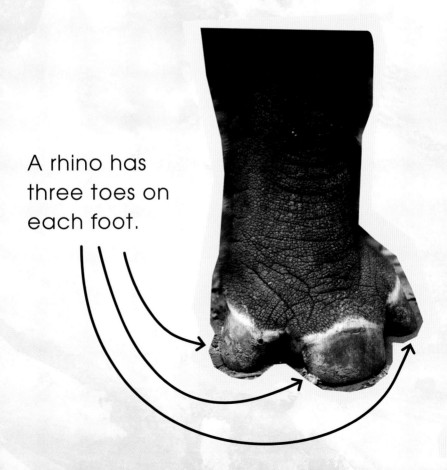

A rhino has three toes on each foot.

what do rhinos eat?

All rhinos eat plants.

Black rhinos use their lips to pull leaves off a tree.

White rhinos have wide mouths.
They use their lips to pull and
tear grass.

Rhinos are big animals.
They eat a lot of food every day!

In the mud

Rhinos live in hot places.

They like to **wallow** in the mud. The mud cools their skin.

Little bugs cannot bite or sting the rhino when it has a coat of mud!

Indian rhinos love the water. They are good swimmers.

Look, listen, smell

All rhinos have small eyes. They do not see well.

A rhino has big ears. It can hear very quiet sounds.

The rhino sniffs the ground. It has a very good **sense** of smell.

The holes in a rhino's nose are called nostrils.

Jungle life

A Sumatran rhino sleeps during the day.

The sun goes down and the rhino gets up.

It is time to eat leaves, **twigs**, and fruit.

The rhino is shy.
If it hears a sound it will run and hide in the jungle.

There are tall trees in the jungle, and places to hide.

Herds

Most rhinos live alone. Some white rhinos live in small **herds**.

There are females and young rhinos in the herd.

Females rub their horns together.

A nose-rub is a good way to say "hi"!

Baby rhinos

A baby rhino is a called a **calf**.

A rhino mom has one calf at a time.

The calf grows inside its mom. It is small when it is born.

Mom feeds her baby with milk and it grows bigger.

A baby rhino stays with Mom for up to four years.

Growing up

Mom looks after her calf and keeps it safe.

If another animal comes near her calf, Mom chases it away.

Mom shows her calf the the best plants to eat.

The Mom and her calf talk to each other. They make many odd sounds!

Rhinos snort, whistle, and honk to talk!

on the move

Rhinos walk slowly. They look, listen, and sniff as they move.

Rhinos use their horns to move through the jungle.

When a rhino is scared it can run fast!

Sumatran rhinos are the fastest rhinos. They can run up a hill or climb out of a river.

Danger!

Rhinos are in danger. People hunt them for their horns.

Hunters sell the horns.

People also cut down the trees in the rainforest. Now some Javan rhinos have no home.

There are only about 50 Javan rhinos left in the world.

Saving rhinos

Long ago, many Sumatran rhinos lived in the jungle.

Now, they are very **rare**. There are about 250 of them left.

Rhinos may go **extinct**.
When they are extinct,
there will be no rhinos left.

Scientists look after rhinos in safe places. They work hard to **protect** them.

Glossary

Calf
A young animal.

Extinct
When all the animals of one type have died out.

Grassland
An area of land where few trees grow but there is lots of grass.

Herd
A group of animals.

Hunter
A person who finds animals and kills them.

Jungle
A forest that grows in a hot place.

Protect
To take care of something and keep it safe.

Rainforest
A forest that grows where it's hot and it rains most days.

Rare
An animal is rare when there are not many of them.

Sense
Animals have five main senses: seeing, hearing, touching, tasting, and smelling.

Twig
A thin part of a branch.

Wallow
Roll around or lie in water, or in mud.

Woods
A place where many trees grow.

Thanks for reading!